AF822782

Who Am I?
Forest Animals

Written by Read With You
Center for Language Research and Development

The characters and events portrayed in this book are fictitious. Any similarity to real persons, living or dead, is coincidental and not intended by the author.

Published by Read With You Publishing

Designed by Read With You Center for Language Research and Development

ISBN-13: 978-1-944710-85-9
ISBN-10: 1-944710-85-X

Printed in the United States of America.

I open up my quills when I'm scared.

Who am I?

porcupine

I can turn my ears in any direction without moving my head.

Who am I?

deer

I’m active at night.

Who am I?

raccoon

I can wrap my tongue around my head.

Who am I?

woodpecker

My fur is soft and glossy.

Who am I?

mink

I have a long, hard, and straight snout.

Who am I?

wild boar

My sharp teeth never stop growing.

Who am I?

squirrel

I live and hunt in a pack.

Who am I?

wolf

Porcupine

Deer

Raccoon

Woodpecker

I am a porcupine.

I am a deer.

I am a raccoon.

I am a woodpecker.

Mink

Wild Boar

Squirrel

Wolf

I am a mink.

I am a wild boar.

I am a squirrel.

I am a wolf.

We are forest animals.

A forest is a land area with many trees.

Made in United States
North Haven, CT
18 December 2024

63129402R00015